LITTLE BOOK
OF
COWS

WEIDENFELD & NICOLSON
LONDON

COWS

MAJESTIC AND MYSTERIOUS

I think
   There's a summer ocean liner in cows -
Majestic and far off,
   With a quiet mysterious delight,
Fading through the blue afternoon.

*A*nd there's a ruined holy city
   In a herd of lying down, cud-chewing cows -
Noses raised, eyes nearly closed
   They are fragments of temples - even their outlines
Still at an angle unearthly.
   As if a ray from heaven still rested across their brows,
As if they felt it, a last ray.

WHAT IS THE TRUTH?
Ted Hughes

IN THE MEADOWS AT CANTERBURY
William Sidney Cooper 19th century

COWS

FEEDING TIME *Walter Hunt* 1861 - 1941

Mistaking The Moon

T HE COW standing erect was of
the Devon breed, and was
encased in a tight warm hide of
rich Indian red, as absolutely
uniform from eyes to tail as if the
animal had been dipped in a dye of
that colour, her long back being
mathematically level. The other
was spotted, grey and white. Beside
her Oak now noticed a little calf
about a day old, looking idiotically
at the two women, which showed
that it had not long been
accustomed to the phenomenon of
eyesight, and often turning to the
lantern, which it apparently
mistook for the moon, inherited
instinct having as yet had little time
for correction by experience.

FAR FROM THE
MADDING CROWD
Thomas Hardy 1840-1928

COWS

THE PIPER

There was a piper had a cow
And he had nought to give her.
He pulled out his pipes and played her a tune,
And bade the cow consider.

The cow considered very well
And gave the piper a penny,
And bade him play the other tune,
'Corn rigs are bonny'.

TRADITIONAL

IN AN ORCHARD
Henry William Banks Davis
1833 - 1914

To a Cow

They took your calf away last night,
　　So that is why you moo
And all the beasts in sympathy
　　Mourn from the field with you!

Commiseration flows from me
　　It flows from every part
As lying still I hear that low
　　From out your bovine heart.

Maternal anguish racks your frame
　　And yet you cannot weep,
Just bellow sadly to the stars -
　　But please, I want some sleep.

M. James

COWS

THE BELLMAN *Samuel Palmer* 1 8 0 5 - 1 8 8 1

C O W S

Formidable Servants

n both sides of the road, in big dusty fields, farmers were preparing for next spring. Every fifty yards a yoke of great-necked stolid oxen were patiently haling at the plough. I saw one of these mild, formidable servants of the glebe, who took a sudden interest in Modestine and me. The furrow down which he was journeying lay at an angle to the road, and his head was solidly fixed to the yoke like those of caryatides below a ponderous cornice; but he screwed round his big honest eyes and followed us with a ruminating look, until his master bade him turn the plough and proceed to reascend the field. From all these furrowing ploughshares, from the feet of oxen, from a labourer here and there who was breaking the dry clods with a hoe, the wind carried away a thin dust like so much smoke.

TRAVELS WITH A DONKEY IN CEVENNES
Robert Louis Stevenson 1850-1894

COWS

THE WATERING PLACE

Frederick William Hulme 1816-1884

COWS

TENDING THE CALVES
Richard Ansdell 1815-1885

COWS

Twins

Born in the spring at dead of frosty night,
The moon on cowshed thatch and sleeping house.
Thin cries in darkness, with the lamp's small light
Falling on upturned pail, sharp watching mouse.

A few days old to stand on glassy feet,
Nuzzling together in the littered hay,
Soft liquid eyes at peace, bleat after bleat,
They tug at milky udders all the day.

Let loose at last from smells of raftered home,
Inherit fields of grass and pasture flowers,
A continent of green in which to roam,
Companions of the dawn and twilight showers.

We see them grazing from behind the fence,
And wonder why such frisky calves must grow
To dawdling cows, so stolid and immense,
Who only knew the world a year ago.

Leonard Clark

C O W S

I Had a Little Cow

I had a little cow: to save her,
I turned her into the meadow to graze her;
There came a heavy storm of rain,
And drove the little cow home again.
The church doors they stood open,
And there the little cow cropen;
The bell ropes they were made of hay,
And the little cow ate them all away;
The sexton came to toll the bell,
And pushed the little cow into the well!

TRADITIONAL

THE SHINING RIVER
Samuel Palmer 1805-1881

The Cattle Camp

After an evening of ghost stories, a creaking door is enough to set teeth chattering; and after an evening of cattle-camp yarns, told in a cattle camp, a snapping twig is enough to set hair lifting; and just as the most fitting place for ghost stories is an old ruined castle, full of eerie noises, so there is no place more suited to cattle-camp yarns than a cattle camp. They need the reality of the camp-fire, the litter of camp baggage, the rumbling mob of shadowy cattle near at hand.

WE OF THE NEVER NEVER
Mrs Aeneas Gunn 1870-1961

THE COW

The friendly cow, all red and white,
I love with all my heart:
She gives me cream with all her might,
To eat with apple-tart.

She wanders lowing here and there,
And yet she cannot stray,
All in the pleasant open air,
The pleasant light of day;

And blown by all the winds that pass
And wet with all the showers,
She walks among the meadow grass
And eats the meadow flowers.

A CHILD'S GARDEN OF VERSES
Robert Louis Stevenson 1850 - 1894

LANDSCAPE WITH COWS BY A RIVER

Henry Moore 1 8 3 1 - 1 8 9 5

SPRING

n the first of May the cows left their winter quarters in the cowhouses, and were turned out to graze in the fields. That was a day to remember! Becky put her hands on her hips and shouted with laughter at their antics as they came pushing, tumbling through the gate and galloped wildly up and down the hills, with outstretching tails and tossing horns. They flung their heads back and blorted, they stamped their feet on the cool soft earth, they leapt like young lambs and danced with their unwieldy bodies on their slender legs.

Cows that had long been jealous attacked each other with curved horns, and the farmer and Dan stood ready with forks and sticks to prevent any harm. They raised their noses in the air and sniffed the smells of spring, and they ran to the streams and water-troughs, trampling the clear fresh water, drinking deeply with noisy gulps. They explored their old haunts, rubbed their flanks against their favourite stumps and railings, scratched their heads, polished their horns, and then settled down to eat the young, short, sweet grass.

THE COUNTRY CHILD *Alison Uttley* 1884-1976

COWS

The Calf

You may have seen, in road or street,
   At times, when passing by,
A creature with bewildered bleat
   Behind a milcher's tail, whose feet
Went pit-pat. That was I.

     Whether we are of Devon kind,
       Shorthorns or Herefords,
     We are in general of one mind
       That in the human race we find
     Our masters and our lords.

When grown-up (if they let me live)
   And in a dairy-home,
I may less wonder and misgive
   Than now, and get contemplative,
And never wish to roam.

     And in some fair stream, taking sips,
       May stand through summer noons,
     With water dribbling from my lips
       And rising halfway to my hips,
     And babbling pleasant tunes.

Thomas Hardy 1 8 4 0 - 1 9 2 8

CLOSE FRIENDS

George William Horlor 19th century

COWS

A PEASANT HARNESSING A BULL
Eugene Verboeckhovn 1799-1881

COWS

A Farmer's Boy

✴

They strolled down the lane together,
The sky was studded with stars -
They reached the gate in silence
And he lifted down the bars -
She neither smiled nor thanked him
Because she knew not how:
For he was just a farmer's boy
And she was a Jersey cow.

TRADITIONAL

COWS

THE OX

Why should I pause, poor beast, to praise
    Thy back so red, thy sides so white;
And on thy brow those curls in which
    Thy mournful eyes take no delight?

I dare not make fast friends with kine,
    Nor sheep, nor fowl that cannot fly;
For they live not for Nature's voice,
    Since 'tis man's will when they must die.

COWS

So, if I call thee some pet name,
    And give thee of my care to-day,
Where wilt thou be to-morrow morn,
    When I turn curious eyes thy way?

Nay, I'll not miss what I'll not find,
    And I'll find no fond cares for thee;
So take away those great sad eyes
    That stare across yon fence at me.

See you that Robin, by himself,
    Perched on that leafless apple branch,
His breast like one red apple left -
    The last and best of all - by chance?

If I do but give heed to him,
    He will come daily to my door;
And 'tis the will of God, not Man,
    When Robin Redbreast comes no more.

William Henry Davies 1871-1940

COWS

MILKING

Cushy cow, bonny, let down
thy milk,
And I will give thee a gown
of silk;
A gown of silk and a silver tee,
If though wilt let down thy
milk for me.

TRADITIONAL

THE MILKMAID *Julien Dupre* 1851-1910

COWS

If I Might Be An Ox

*I*f I might be an ox,
   An ox, a beautiful ox,
   Beautiful but stubborn:
   The merchant would buy me,
   Would buy me and slaughter me,
   Would spread my skin,
   Would bring me to the market,
   The coarse woman would bargain for me,
   The beautiful girl would buy me,
   She would crush perfumes for me,
   I would spend the night rolled up around her.
   Her husband would say: 'It is a dead skin'.
   But I would have her love.

TRADITIONAL TRIBAL
SONG FROM ETHIOPIA

OX *Thomas Roebuck* 19th century

COWS

When The Cows Come Home

When the cows come home
    the milk is coming;
Honey's made while bees
    are humming;
Duck and drake on the
    rushy lake,
And the deer live safe in the
    breezy brake;
And timid, funny, pert
    little bunny
Winks his nose, and sits
    all sunny.

Christina Rossetti 1830-1894

COWS

GLEANERS RETURNING

James Edwin Meadows 1828-1888

The Old Woman's Three Cows

There was an old woman had three cows,
Rosy and Colin and Dun.
Rosy and Colin were sold at the fair,
And Dun broke her heart in a fit of despair,
So there was an end of her three cows,
Rosy and Colin and Dun.

TRADITIONAL

THE PATH BY THE WATER LANE
Myles Foster 1825-1899

WHERE THE
WILLOWS GROW
Thomas Sidney Cooper
1803 - 1902

COWS

The Gracious and Gentle Thing

*T*he three young heifers were at summer supper
In the cowpen munching new-mown hay,
Their eyes suffused with sweetness of red clover,
It was no time to pass the time of day.
Their chins went side to side, their cheeks were bulging
Indecorously, and they were eating more;
I was a stranger, I had no introduction,
They had never laid eyes on me before.

Yet when I patted each young lady's sleekness,
Each young lady's lips grew bland and still,
She left the hay that sweetened the whole evening
And beamed on me with eyes deep with good will.
She kissed my hand where it lay on the fence-rail
And breathed her sweetness in my smiling face;
She left her supper, turned her slender beauty
Instantly to practice of good grace.

I stood there below the azure evening
With miles of tender thrushes all around
And thought how up and down the land I never
So natural a courtesy had found
As this night in a barnyard with three heifers.
The gracious and the gentle thing to do,
With never any lesson in good manners,
These innocent and courteous creatures knew.

Robert Tristram Coffin

COWS

RIBYSWELL, LINCOLN RED BULLOCK early 20th century

COWS

The Best Beast

The Best Beast of the Show
Is fat
He goes by the lift
They all do that . . .

Is he not fat?
Is he not fit?
Now in a crown he walks
To the lift . . .

I touched his hide
I touched the root of his horns
The breath of the Beast
Came in low moans.

THE BEST BEAST
OF THE FAT STOCK SHOW
AT EARL'S COURT
Stevie Smith 1902-1971

The Character of Cows

here were the extroverts. They were always first into the parlour, took the lead in duffing up the dog or any deer that might be foolish enough to visit their field and were through any open gates like a flash, to career through the countryside creating as much mayhem as possible.

The next largest group were the morons. Just as people can vary in intelligence and dynamism, so can cows. However, a dozy cow is a very stupid animal indeed. The typical moron was that cow of the popular image, slow, sedate, plodding through life as if her batteries were running down.

ANY FOOL CAN BE A DAIRY FARMER
James Robertson

CATTLE ON COAST
Henry William Banks Davis 1833-1914

COWS

THE MAGNIFICENT BULL

My bull is white like the silverfish in the river,
White like the shimmering crane bird on the river bank,
White like fresh milk!
His roar is like thunder to the Turkish cannon on the
    steep shore.

My bull is dark like the rain cloud in the storm.
He is like summer and winter.
Half of him is dark like the storm cloud
Half of him is light like sunshine.
His back shines like the morning star.
His brow is red like the back of the hornbill.
His forehead is like a flag, calling the people from a distance.

He resembles the rainbow.
I will water him at the river,
With my spear I shall drive my enemies.
Let them water their herds at the well;
The river belongs to me and my bull.
Drink, my bull, from the river; I am here to guard
    you with my spear.

TRADITIONAL *Dinka Tribe, Africa*

COWS

CANTERBURY MEADOWS

Thomas Sidney Cooper 1803-1902

❀

COWS

HOMEWARD BOUND

Joseph Horlor early 20th century

The Lost Cow

Simon Brodie
        had a cow;
He lost his cow and
        could not find her;
When he had done what
        man could do
The cow came home
        and her tail behind her.

EDWARD SILL
1841 - 1887

OFF TO PASTURE

Thomas Sidney Cooper

1803 - 1902

COWS

The Cow and the Ass

B ESIDE a green meadow a stream used to flow,
So clear, you might see the white pebbles below;
To this cooling brook, the warm cattle would stray,
To stand in the shade on a hot summer's day.

A cow quite oppressed by the heat of the sun,
Came here to refresh as she often had done;
And, standing quite still, stooping over the stream
Was musing, perhaps - or perhaps she might dream.

But soon a brown ass of respectable look
Came trotting up also to taste of the brook,
And to nibble a few of the daisies and grass.
'How d'ye do?' said the cow. 'How d'ye do?' said the ass.

'Take a seat!' said the cow, gently waving her hand.
'By no means, dear madam,' said he, 'while you stand!'
Then, stooping to drink, with a very low bow,
'Ma'am, your health!' said the ass.
'Thank you, sir,' said the cow.

Jane Taylor 1783-1824 and Ann Taylor 1782-1866

COWS

MY LITTLE COW

I had a little cow,
   Hey diddle, ho diddle!
I had a little cow,
   And I drove it to the stall;
Hey diddle, ho diddle!
   And there's my song all.

TRADITIONAL

COWS

COWS WITH A HERDSMAN

English School 19th century

CATTLE WATERING NEAR YORK
Charles Shayer 19th century

COWS

THINKING THOUGHTS

Between the open door
And the trees two calves
  were wading in the pond,
Grazing the water here
  and there and thinking,
Sipping and thinking,
  both happily, neither long.
The water wrinkled, but
  they sipped and thought,
As careless of the wind
  as it of us.
'Look at those calves.
  Hark at the trees again.'

UP IN THE WIND
Edward Thomas 1878-1917

THE OX
PUT OUT TO GRASS

Alcon hasn't led off to
   the red axe
Worn out by coulter and
   by age his hard-worked ox:
Its labour honoured, in
   the deep grass it now
Happily lows its freedom
   from the plough.

Adaeus the Macedonian
4th century BC

PORTRAIT OF A SHORTHORN *W. Shayer* 19th century

The Little Miracle

I grasped the calf by its fore legs and pulled it up to its mother's head. The cow was stretched out on her side, her head extended wearily along the rough floor. Her ribs heaved, her eyes were almost closed; she looked past caring about anything. Then she felt the calf's body against her face and there was a transformation; her eyes opened wide and her muzzle began a snuffling exploration of the new object. Her interest grew with every sniff and she struggled on to her chest. Then she began to lick methodically. Nature provides the perfect stimulant massage for a time like this and the little creature arched his back as the coarse papillae on the tongue dragged along his skin. Within a minute he was shaking his head and trying to sit up.

I grinned. This was the bit I liked. The little miracle. I felt it was something that would never grow stale no matter how often I saw it.

IF ONLY THEY COULD TALK
James Herriot

COWS

THE ORPHAN
Walter Hunt 1861-1941

Acknowledgements

Designed and edited by
THE BRIDGEWATER BOOK COMPANY
Words and Pictures chosen by
RHODA NOTTRIDGE
Typesetting by VANESSA GOOD
Printed in Italy

*The publishers wish to thank the following
for the use of pictures:*
THE BRIDGEMAN ART LIBRARY:
front cover and pages 9, 18, 31, 53;
E.T. ARCHIVE: pages 16, 38;
FINE ART PHOTOGRAPHS: back cover and
pages 3, 4, 7, 11, 12, 15, 20, 23, 24, 26, 29,
33, 35, 36, 41, 43, 44, 46, 49, 50, 55.

*The publishers gratefully acknowledge permission
to reproduce the following material in this book:*
p.2 *Extract from What is the Truth?* by Ted
Hughes by permission Faber & Faber Ltd.

p.13 *Twins* by Leonard Clark by permission
Robert Clark.
p.17 *We of the Never Never* by Mrs Aeneas
Gunn by permission A.P.Watt Ltd. on behalf
of Peta Paine Nominees Ltd.
p.21 *The Country Child* by Alison Uttley
by permission Faber & Faber Ltd.
p.27 *The Ox* by W H Davies; The Estate of the
W H Davies and Jonathan Cape.
p.39 Extract from *The Best Beast Of The Fat
Stock Show At Earl's Court* in Stevie Smith:
Collected Poems. Copyright ©1982
by Stevie Smith.
p.40 *Any Fool Can Be A Dairy Farmer* by
James Robertson (Farming Press, Wharfedale
Road, Ipswich UK).
p.42 Dinka tribe poem in *Pet Poems* edited by
Robert Fisher, Faber & Faber, 1989.
p.51 Extract from *Up In The Wind* by Edward
Thomas courtesy Myfanwy Thomas and Faber
& Faber Ltd.
p.54 *If Only They Could Talk* by James Herriot
by permission David Higham Associates and
St Martins Press (USA).

*Every effort has been made to trace all copyright
holders and obtain permissions. The editor and
publishers sincerely apologise for any inadvertent
errors or omissions and will be
happy to correct them in any
future edition.*